YOUR KNOWLEDGE HAS VALUE

- We will publish your bachelor's and
  master's thesis, essays and papers

- Your own eBook and book -
  sold worldwide in all relevant shops

- Earn money with each sale

Upload your text at www.GRIN.com
and publish for free

John Peter

Red phosphors for W-LED applications

GRIN Publishing

GRIN - Your knowledge has value

Since its foundation in 1998, GRIN has specialized in publishing academic texts by students, college teachers and other academics as e-book and printed book. The website www.grin.com is an ideal platform for presenting term papers, final papers, scientific essays, dissertations and specialist books.

Visit us on the internet:

http://www.grin.com/

http://www.facebook.com/grincom

http://www.twitter.com/grin_com

A Novel $Li_3Ba_2Gd_3(MoO_4)_8:RE^{3+}$ ($RE= Pr^{3+}, Sm^{3+}$) red phosphors for W-LED applications.

ANTHUVAN JOHN PETER
Department of Physics, St. Anne's College of Engineering and Technology, Panruti, Tamilnadu, India

ABTRACT

A series of $Li_3Ba_2Gd_{3-x}Pr_x(MoO_4)_8$ (x = 0.01, 0.03, 0.05, 0.07and 0.09 mol) and $Li_3Ba_2Gd_{3-x}Sm_x(MoO_4)_8$ (x = 0.02, 0.04, 0.06, 0.08 and 0.10 mol) red phosphors were synthesized by conventional solid state reaction method. For 7 mol % of Pr^{3+} and for 8 mol% of Sm^{3+} concentration, the phosphors exhibit a strong excitation peak at 450 nm (blue light) and 401 nm (near UV) indicating an intense absorption. The PL emission spectra of $Li_3Ba_2Gd_3(MoO_4)_8:Pr^{3+}$ and $Li_3Ba_2Gd_3(MoO_4)_8:Sm^{3+}$ phosphors showed an intense peak at 645 nm (red) which corresponds to $^3P_0\rightarrow{}^3F_2$ transition of Pr^{3+} and at 604 nm (red) which corresponds to$^4G_{5/2}\rightarrow{}^6H_{7/2}$ transition of Sm^{3+}. The optimal doping concentration of Pr^{3+} doped $Li_3Ba_2Gd_3(MoO_4)_8$ was measured to be 0.07 mol and that for Sm^{3+} doped $Li_3Ba_2Gd_3(MoO_4)_8$ was 0.08 mol. Our results indicate that $Li_3Ba_2Gd_3(MoO_4)_8$ is a potential host material for Pr^{3+} and Sm^{3+} and $Li_3Ba_2Gd_3(MoO_4):RE^{3+}$ ($RE^{3+} = Pr^{3+}, Sm^{3+}$) is capable of converting the blue /near UV emission of a light-emitting diode into red light.

Content

1. Introduction

Currently, trivalent rare-earth-ion-activated molybdate based phosphors have fascinated great attention for solid-state lighting applications by virtue of their long lifetimes, and efficient luminescence property. The rare-earth ions are represented by a partly filled 4f shell that is completely shielded by $5s^2$ and $5p^6$ orbitals. Therefore, emission transitions provide sharp intense lines in the optical spectra [1, 2]. The use of rare-earth element-based phosphor, based on ''line-type'' f–f transitions, can narrow the emissions to the visible range, resulting in high efficiency and a high-lumen equivalence. In the recent years, a flourishing care is concentrated on $Li_3Ba_2Gd_{3-x}(MoO_4)_8$ host matrix for luminescent ions in the interest of their excellent chemical and thermal stability and favourable luminescence characteristics compared to the sulfide- and nitride-based materials. Moreover, these are environmentally friendly as no toxic gases like sulphide are given out. $Li_3Ba_2Gd_{3-x}(MoO_4)_8$ occur in monoclinic crystal system with space group $C2/c$ in a disordered structure [3].

They are remarkable host matrices for luminescent ions due to their unique structure, good chemical stability, low temperature preparation, superior luminescence features and low cost raw materials.

They have also brought a lot of significance for their unique characteristics like ferroelectricity, laser hosts, phosphors and catalysis [4, 5]. The investigations on the luminescence properties of rare earth doped $Li_3Ba_2Gd_{3-x}(MoO_4)_8$ are scanty. Among these, most of the earlier studies carried on these materials in the past few years specifically focused luminescence properties of the single crystals of $Li_3Ba_2Gd_3 (MoO_4)_8$ and $Li_3Ba_2Gd_3 (MoO_4)_8$ activated with various lanthanide ions [6]. Recently Song et al [7] studied the crystal structure and optical properties of $Li_3Ba_2Gd_3 (MoO_4)_8:Tm^{3+}$. The synthesis and characterisation of single and pure phase $Li_3Ba_2Gd_3 (MoO_4)_8: Eu^{3+}$ powders were analysed by Chang et al [8]. Though Eu and Tm doped $Li_3Ba_2Gd_3 (MoO_4)_8$ phosphors have been reported earlier, Pr^{3+} and Sm^{3+} activated $Li_3Ba_2Gd_3 (MoO_4)_8$ materials are not investigated so far.

Among the many rare earth elements, Pr^{3+} and Sm^{3+}-activated phosphor materials with full color luminescence have drawn considerable attention because the Pr^{3+} ions exhibits a number of distinct emissions depending on the host in which they are fused. For example, Pr^{3+} ions with $4f^2$ electronic configuration exhibit red (from the $^1D_2$ level), green (from the $^3P_0$ level), blue (from the $^1S_0$ level) and ultraviolet (from the 4f–5d state) [9] and Sm^{3+} ions with $4f^5$ electronic configuration exhibit a strong orange–red fluorescence in the visible region [10, 22, 23, 24, 25]. Recently, Molybdates activated with Pr^{3+} and Sm^{3+} are extensively investigated as materials for photonics and solid state lighting devices. Xu et al doped Ag^+ and Y^{3+} in $Li La (MoO_4)_2:Pr^{3+}$ phosphor and finds that the emission intensity drastically increased and the CIE chromaticity coordinates of the as-prepared phosphors are tuned from

(0.639, 0.351) to (0.653, 0.337) [11]. E. Cavalli and his co-worker studied the emission dynamics of the $CaMoO_4$: Pr system under different experimental conditions for white light emission[12]. Yang and his co workers reported the successful synthesis of highly luminescent hollow porous microsphere of $Ba_{0.995}MoO_4$: 0.005Pr phosphor in comparison with the commercially used red phosphor $Y_2O_2S:Eu^{3+}$ [13]. Recently, Lin and his co workers have prepared $SrMoO_4:Sm^{3+}$ phosphors by conventional solid state reaction method at 950^0C and studied the effect of charge compensation on the fluorescence process of Sm^{3+} ion. It is observed that the emission intensity of $Sr_{0.96}MoO_4:0.02Sm^{3+}$, $0.02Na^+$ phosphor was slightly lower than the commercial red $Sr_2Si_5N_8:Eu^{2+}$ phosphor the emission [14].

Taking into account the above investigations, in the present work. We mainly focus on the synthesis and luminescent characteristics of $Li_3Ba_2Gd_{3-x}Pr_x$ $(MoO_4)_8$ (x = 0.01, 0.03, 0.05, 0.07and 0.09 mol) and $Li_3Ba_2Gd_{3-x}Sm_x$ $(MoO_4)_8$ (x = 0.02, 0.04, 0.06, 0.08 and 0.10 mol) red phosphor ceramics by a mechanochemically assisted high temperature solid state reaction method. Furthermore, we investigated the structure, morphology, particle size and photoluminescence (PL) and life time properties of the resulting samples in detail. Finally, from the analysis of the CIE diagram, we propose that the obtained nanoparticles of Pr and Sm doped $Li_3Ba_2Gd_3$ $(MoO_4)_8$ phosphors are promising red emitting components which could be used in White LEDs.

2. Experimental Method

2.1 Preparation of Phosphors

The efficient red phosphors of $Li_3Ba_2Gd_{3-x}Pr_x$ $(MoO_4)_8$ (x = 0.01, 0.03, 0.05, 0.07, 0.09) and $Li_3Ba_2Gd_{3-x}Sm_x$ $(MoO_4)_8$ (x = 0.02, 0.04, 0.06, 0.08 and 0.10 mol)   were synthesized by mechanochemically assisted high temperature solid state reaction method by using AR grade Li_2CO_3, $BaCO_3$, Gd_2O_3 (99.99% Aldrich), Pr_2O_3 (99.99% Aldrich),Sm_2O_3(99.99% Aldrich) and MoO_3 (99.5% Aldrich) as starting materials without any further purification. The relevant stoichiometric mixtures of starting materials were mixed and were milled for a period of three hours in a high energy planetary ball mill Pulverisette 7 (FRITSCH). Milling was carried out in two grinding vials of 15 ccm volume containing balls with diameter of 12 mm. Both the container and balls were made of tungsten carbide material. The milling speed was fixed at 350 revolutions per minute (rpm).The powder samples were calcined in air at 800 °C for 5h with the heating rate of 5K min^{-1} to form polycrystalline $Li_3Ba_2Gd_{3-x}Pr_x(MoO_4)_8$ and $Li_3Ba_2Gd_{3-x}Sm_x(MoO_4)_8$.

2.2 Characterization

X – ray powder diffraction (XRD) analysis was out using Pan Analytical X'pert pro x- ray diffractometer with Cu K-alpha radiation (λ = 1.5406 Å) at a scanning rate of $0.02°$ per second. The XRD patterns were obtained in the range of $0° \leq 2\theta \leq 70°$ and were compared with the JCPDS data. Fourier Transform infrared spectroscopy (FTIR) measurements were carried out in the wavelength range of 400–4000 cm^{-1} with a Nicolet 6700 FTIR equipped with a deuterated triglycine sulfate detector. The particle size and morphology of the powder samples were inspected by using scanning electronic microscope (SEM Philips XL30). The centrifugal particle size analyzer of Shimadzu SA-CP3 was used to observe the distribution and size of the particles. The measurements of PL and photoluminescence excitation (PLE) spectra were performed by a Jobin Yuvon Flurolog-3-11 Spectroflurometer at room temperature with 450W xenon lamp was used as the excitation source (200-700 nm). The phosphor powders were slackly positioned on a solid stand and were excited straight. The excitation and emission slit width were set to 3 nm. Optical absorption spectra were recorded using Systronics-2101 UV–visible- double beam spectrophotometer. All spectroscopic measurements of the phosphors were carried out at room temperature.

3. Results and discussion

3.1 XRD and size distribution characterization

A typical powder X-ray diffraction (XRD) pattern of selected samples is represented in Fig. 1. It can be observed that all of the peaks match those of scheelite-type $Li_3Ba_2Gd_3$ $(MoO_4)_8$ (No.JCPDS 77-0830) structure. The strong and narrow peaks indicate a high crystallinity of the as-prepared products, which is beneficial for high luminescence [3].The peaks in the XRD spectra are sharp and intense proving that a highly crystalline single-phase of $Li_3Ba_2Gd_3$ $(MoO_4)_8$ phosphors with the monoclinic structure of space group C2/c had been successfully prepared without any impurity phase and distortion in crystal structure by mechanochemically assisted high temperature solid state reaction. Therefore, the substitution Pr^{3+} ions and Sm^{3+}were incorporated into the $Li_3Ba_2Gd_3$ $(MoO_4)_8$ host lattice without any significant structural change or observed impurity phase. Based on the effective ionic radii of cations reported by Shannon [15], the ionic radii of Pr^{3+} (r = 1.126 Å, CN = 8, where CN is the coordination number of the metal ions) and Sm^{3+} (r = 1.079 Å, CN = 8, where CN is the coordination number of the metal ions) are closer to that of Gd^{3+} (r = 1.053 Å, CN = 8) when compared to that with Ba^{2+} (r = 1.61Å, CN = 12) and Mo^{6+} (r = 0.59 Å, CN = 6). Therefore Pr^{3+} ions and $Sm^{3+}$are expected to occupy Gd^{3+} sites with eight coordinates preferably. The approximate particle size of the product is calculated from (131) peak according to the Debye-Scherrer's equation: $D = 0.89\lambda\ /\ (\beta\cos\theta)$

where D is the average grain size, λ represents Cu Kα wavelength 0.1542 nm and β is the FWHM of the peak at Bragg angle θ. The estimated results show that the average crystallite size of $Li_3Ba_2Gd_{3\text{-}x}(MoO_4)_8$: x Pr^{3+} (0.01, 0.03, 0.05, 0.07 and 0.09 mol) is about 89 nm and that of $Li_3Ba_2Gd_{3-x}Sm_x(MoO_4)_8$ (x = 0.02, 0.04, 0.06, 0.08 and 0.10 mol) is about 67 nm respectively.

The particle size distribution, EDAX results and SEM image of $Li_3Ba_2Gd_3(MoO_4)_8$:0.07 Pr^{3+} and $Li_3Ba_2Gd_3(MoO_4)_8$:0.08 Sm^{3+} phosphors are shown in Fig. 2 and in Figure.3 respectively. From SEM and particle size distribution data it confirms that the morphologies of $Li_3Ba_2Gd_3(MoO_4)_8$:0.07 Pr^{3+} and $Li_3Ba_2Gd_3(MoO_4)_8$:0.08 Sm^{3+} nano particles appear to be highly crystalline and are slightly agglomerated and the particles show a narrow size distribution, with the average diameter of particles at 85 nm for $Li_3Ba_2Gd_3(MoO_4)_8$:0.07 Pr^{3+} and 65 nm for $Li_3Ba_2Gd_3(MoO_4)_8$:0.08 Sm^{3+} indicating that the particles are fit for the fabrication of solid state lighting devices. [16]

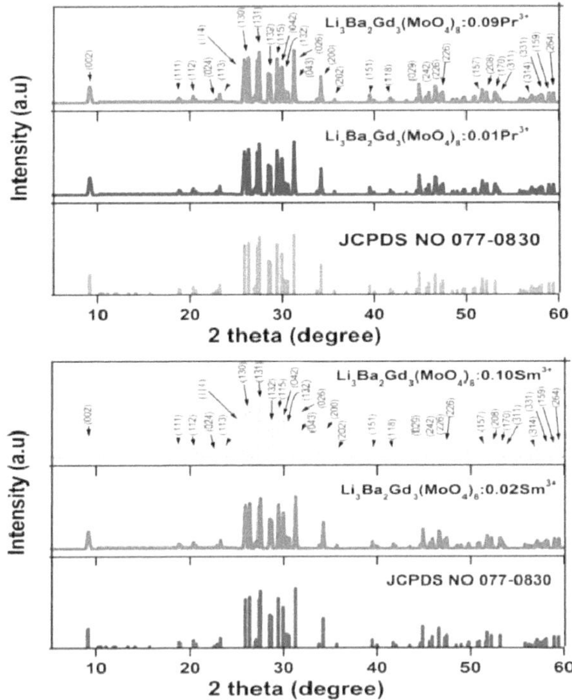

Fig.1. Powder XRD patterns of $Li_3Ba_2Gd_{3-x}$ $(MoO_4)_8:xPr^{3+}$ (x= 0.01and 0.09 mol) and $Li_3Ba_2Gd_{3-x}$ $(MoO_4)_8:xSm^{3+}$ (x= 0.02and 0.10 mol)

The average grain size of the near rounded nano particles is about 82 nm and 68 nm which is consistent with the data of the corresponding XRD patterns. The EDAX of $Li_3Ba_2Gd_3$ $(MoO_4)_8:0.07$ - Pr^{3+} and $Li_3Ba_2Gd_3$ $(MoO_4)_8.0.08$ Sm^{3+} are shown as an example to confirm the presence of the constituent elements. EDAX spectrum shows the respective peaks of the constituent elements such as Lithium (Li), Barium (Ba), Gadolinium (Gd), Oxygen (O), Molybdenum (Mo), Praseodymium (Pr) and samarium (Sm) for EDAX of $Li_3Ba_2Gd_3(MoO_4)_8:0.07$ Pr^{3+} and Gadolinium (Gd) in $Li_3Ba_2Gd_3(MoO_4)_8:0.08$ Sm^{3+} compounds. The stoichiometric proportions of the crystal as well as the EDAX results are in good agreement. The average grain size of the near rounded nano particles is about 82 nm and 68 nm which is consistent with the data of the corresponding XRD patterns.

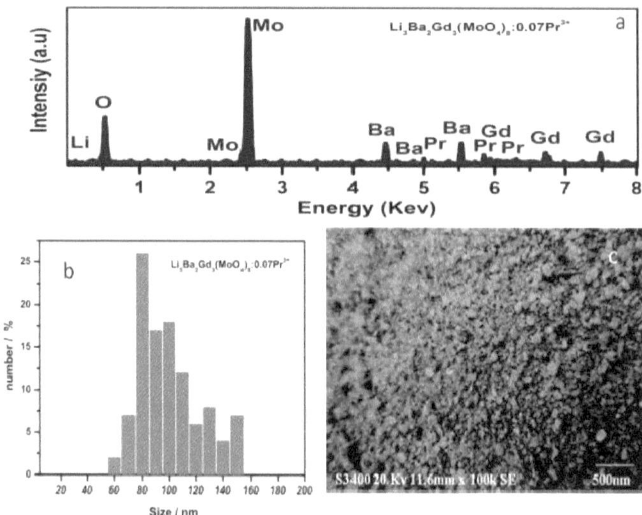

Fig.2. EDAX (a), Particle size distribution (b), and SEM (c) of Li₃Ba₂Gd₃ (MoO₄)₈:0.07 Pr³⁺ Phosphor

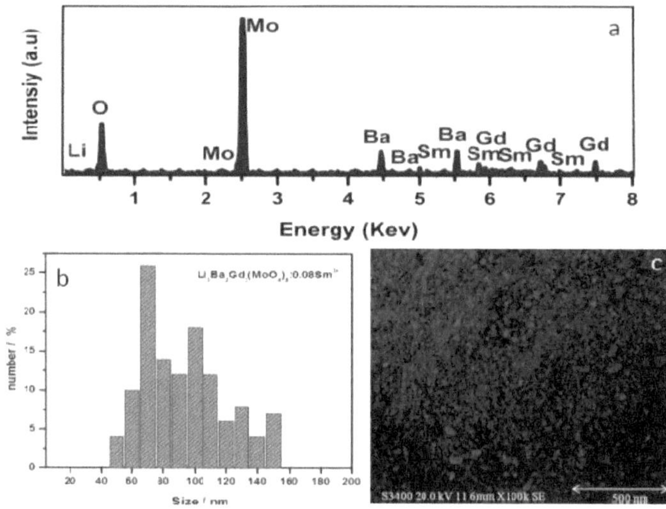

Fig.3. EDAX (a), Particle size distribution (b), and SEM (c) of Li₃Ba₂Gd₃(MoO₄)₈:0.08Sm³⁺ phosphor

3.2 FTIR analysis

The FT-IR spectra of $Li_3Ba_2Gd_3(MoO_4)_8$:0.07 Pr^{3+} and $Li_3Ba_2Gd_3(MoO_4)_8$:0.08 Sm^{3+} red nanophosphors prepared by mechanochemically assisted high temperature solid state reaction is shown in Fig. 4.

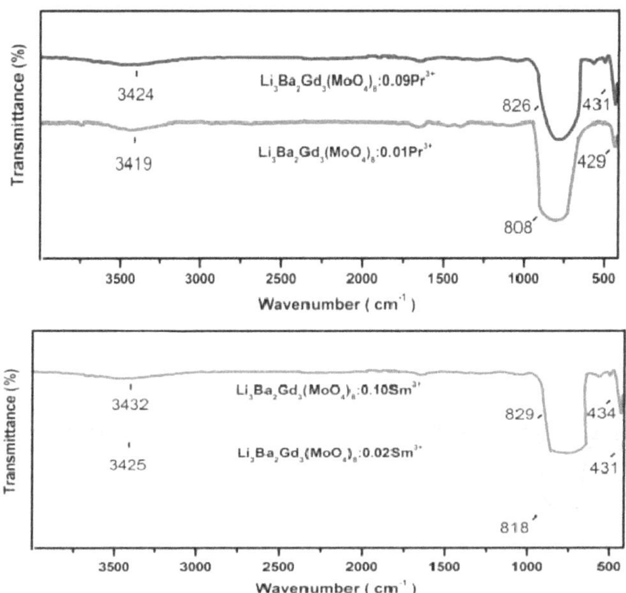

Fig.4. FTIR spectra of $Li_3Ba_2Gd_{3-x}$ $(MoO_4)_8$:xEu^{3+} (0, 0.01 and 0.09 mol) and $Li_3Ba_2Gd_{3-x}$ $(MoO_4)_8$: xSm^{3+} (x= 0.02 and 0.10 mol)

The bands at around 3363 cm^{-1} and at around 1649 cm^{-1} are assigned to O-H stretching vibration and H-O-H bending vibration, respectively[17,18]. These two bands are the characteristic vibrations of water molecules absorbed from air by the sample surface. The strong absorption peaks at around 911.9, 807.1 and 728.0 cm^{-1} are assigned to stretching vibration of O–Mo–O in MoO_4^{2-} tetrahedron. [19]

3.3 Photoluminescence Properties of $Li_3Ba_2Gd_3$ $(MoO_4)_8$: Pr^{3+}:

3.3.1 The excitation and emission spectra of $Li_3Ba_2Gd_3$ $(MoO_4)_8$: Pr^{3+} phosphor

Optical properties of $Li_3Ba_2Gd_3$ $(MoO_4)_8$: Pr^{3+} (x= 0.01, 0.03, 0.05, 0.07 and 0.09 mol) phosphors prepared by mechanochemically assisted solid state reaction method is shown in Fig. 5 and Fig.6. The excitation spectrum is monitored at the emission wavelength of 645 nm. It can be seen clearly that the excitation spectrum consists of a strong and broadband from 250 to 350 nm with a maximum at about 294 nm and this is the charge-transfer band (CTB) ascribed to the host. In addition to the CTB, some sharp absorption peaks are found in the wavelength region of 450–500 nm, resulting from the excitation of the f–f shell transitions of Pr^{3+} ions, and the excitation bands consist of three main peaks located at 453, 474 and 487 with the maximum excitation wavelength at 453 nm which is in well agreement with the blue wavelength of GaN based LED chips. These excitation peaks are respectively attributed to the electronic transitions of $^3H_4 \rightarrow ^3P_2$, $^3H_4 \rightarrow ^3P_1 + ^1I_6$, $^3H_4 \rightarrow ^3P_0$ which are all due to the typical f–f transitions of Pr^{3+} ions.

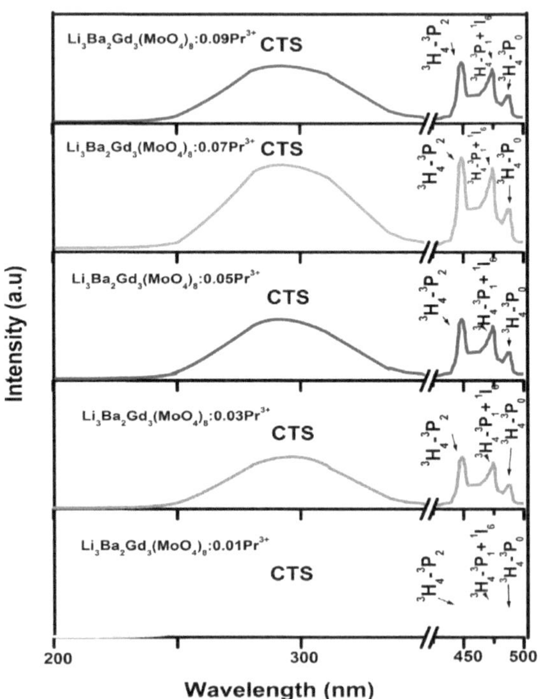

Fig.5. The photoluminescence excitation (λ_{em} = 645 nm) spectra of $Li_3Ba_2Gd_{3-x}$ $(MoO_4)_8$: xPr^{3+} (x= 0.01, 0.03, 0.05, 0.07 and 0.09 mol)

Upon excitation at 453 nm, the obtained emission spectrum exhibits four major emission bands at 594, 618, 645 and 680 nm, corresponding to the $^3P_0 \rightarrow ^3H_5$, $^3P_0 \rightarrow ^3H_6$, $^3P_0 \rightarrow ^3F_2$, $^3P_0 \rightarrow ^3F_3$ typical transitions of Pr^{3+} respectively. The strongest peak appearing at 645 nm is the characteristic emission of Pr^{3+} with $^{3}P_0 \rightarrow {}^{3}F_2$ red emission. It was monitored from the PL spectra that the relative intensity of the Pr^{3+} emission increase as the concentration of the Pr dopant increases and reaches a maximum at x=0.07. With further of addition of Pr^{3+} PL decreases because of the concentration quenching which is caused by the energy transfer between the identical Pr^{3+} ions. This promotes an increase of the non-radiative relaxation and may be responsible for the Pr^{3+} luminescence inhibition in Gd system beyond 7 mol% [19].

3.3.2 The excitation and emission spectra of Li₃Ba₂Gd₃ (MoO₄)₈: Sm³⁺ phosphor

Optical properties of the $Li_3Ba_2Gd_{3-x}Sm_x$ $(MoO_4)_8$ (x = 0.02, 0.04, 0.06, 0.08 and 0.10 mol) phosphors prepared by mechanochemically assisted solid state reaction method is shown in Fig.7 and Fig.8

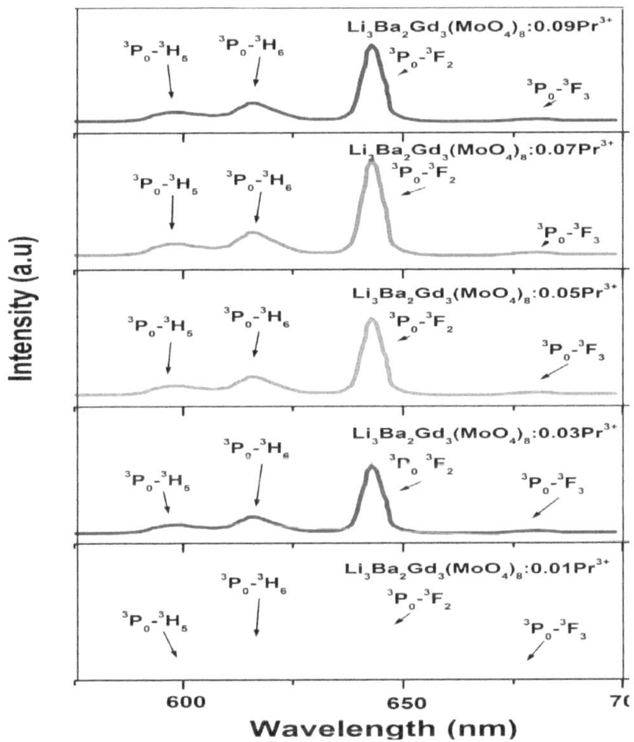

Fig.6. The photoluminescence emission (λex = 450 nm) spectra of Li$_3$Ba$_2$Gd$_{3-x}$ (MoO$_4$)$_8$:xPr^{3+} (x= 0.01, 0.03, 0.05, 0.07 and 0.09 mol)

The excitation spectrum (λ_{em} = 605 nm) transitions consist of a most intense and broad excitation band in the range from 225 nm to 330 nm (CTS) and very sharp peaks with remarkable intensity in the range from 330 nm to 500 nm which corresponds to the typical Sm^{3+} intra – 4f transitions in the host lattice. The strong absorption at 404 nm is attributed to the ^{6}H$_{5/2}$- ^{4}G$_{7/2}$transition of Sm^{3+} which is in good agreement with the near-UV wavelength of GaN based LED chips.

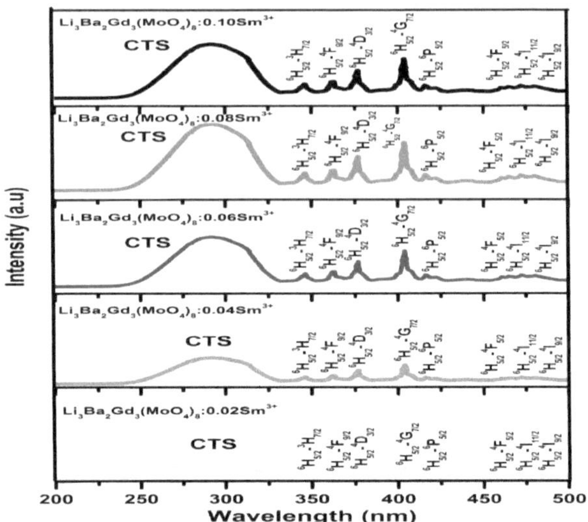

Fig.7. The photoluminescence excitation (λem = 605 nm) spectra of Li$_3$Ba$_2$Gd$_{3-x}$ (MoO$_4$)$_8$:x Sm^{3+} (x= 0.02, 0.04, 0.06, 0.08 and 0.10 mol)

Upon excitation at 404 nm (near UV), the obtained emission spectra exhibits three sharp lines which are associated with the ^{4}G$_{5/2}\rightarrow$ ^{6}H$_J$ (J = 5/2, 7/2,9/2) transitions from the excited Sm^{3+} to the ground state as shown in Fig.8. The main emission peak is the ^{4}G$_{5/2}\rightarrow$ ^{6}H$_{7/2}$ transitions of Sm^{3+} at 605 nm (red). Other transitions located in the range of 530 nm to 750 nm are weak.The optimum concentration of Li$_3$Ba$_2$Gd$_{3-x}$Sm$_x$ (MoO$_4$)$_8$ (x = 0.02, 0.04, 0.06, 0.08 and 0.10 mol) at which maximum emission occurs is found to be 8 mol%.

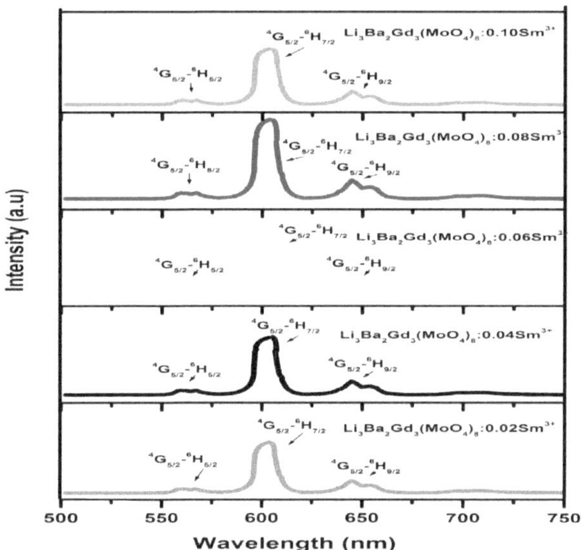

Fig. 8. The photoluminescence emission (λ_{ex} = 404 nm) spectra of Li$_3$Ba$_2$Gd$_{3-x}$ (MoO$_4$)$_8$:xSm^{3+} (x= 0.02, 0.04, 0.06, 0.08 and 0.10 mol)

As the present work is focused for the purpose of red phosphor application in WLED with near UV/blue GaN based chips as excitation sources, the results of the excitation and emission spectra of the present work are compared with that of emission spectra of the conventional red phosphor Y$_2$O$_2$S: Eu^{3+} prepared by solid state reaction method. For this purpose, the commercial red phosphor Y$_2$O$_2$S: Eu^{3+} was prepared by solid state reaction (SSR) method and the luminescence properties were also studied. The red emission spectra of these red phosphors are presented in Fig.9 which reveals that red emission intensity of Li$_3$Ba$_2$Gd$_3$ (MoO$_4$)$_8$:0.07 Pr^{3+} and Li$_3$Ba$_2$Gd$_3$ (MoO$_4$)$_8$: 0.08 Sm^{3+} are comparable with that of commercial Y$_2$O$_2$S: Eu^{3+} red phosphor, indicating that Li$_3$Ba$_2$Gd$_3$ (MoO$_4$)$_8$:0.07 Pr^{3+} and Li$_3$Ba$_2$Gd$_3$ (MoO$_4$)$_8$: 0.08 Sm^{3+} phosphors prepared by mechanochemically assisted high temperature solid state reaction is a promising red phosphor for WLEDs.

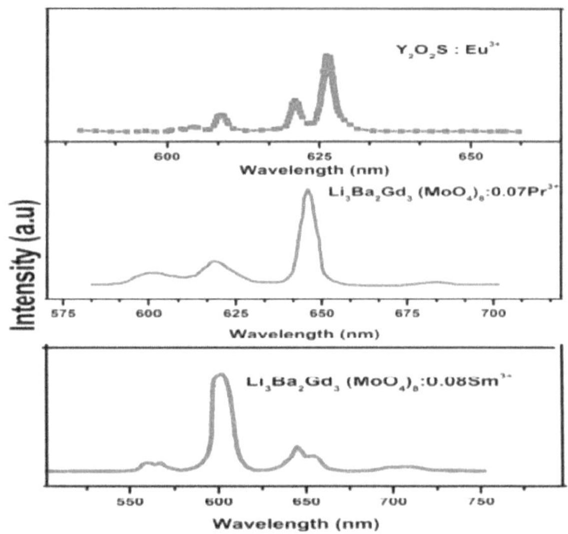

Fig. 9. Comparison of emission spectra of Li₃Ba₂Gd₃ (MoO₄)₈:0.07 Pr^{3+}, Li₃Ba₂Gd₃ (MoO₄)₈:0.08Sm^{3+} and Y₂O₂S: Eu^{3+} phosphors.

4. Conclusion:

Highly luminescent red nanophosphors of Li₃Ba₂Gd₃ (MoO₄)₈:RE^{3+} (RE= Pr^{3+}, Sm^{3+}), with pure single phase and with different gain sizes, were successfully prepared by mechanochemically assisted high temperature solid state reaction method. Upon 450 nm blue excitation, the Li₃Ba₂Gd₃ (MoO₄)₈:0.07 - Pr^{3+} phosphor showed strong red emission lines at 645 nm. The optimum doping concentration of Pr^{3+} content in Li₃Ba₂Gd₃ (MoO₄)₈ for the enhanced red emission is found to be 7mol%. Upon 404 nm near UV excitation, the Li₃Ba₂Gd₃ (MoO₄)₈:0.08Sm^{3+} phosphor showed intense red emission lines at 605 nm. The optimum doping concentration of Sm^{3+} content in Li₃Ba₂Gd₃ (MoO₄)₈ for the enhanced red emission are found to be 8mol%. The red emission intensity of Li₃Ba₂Gd₃ (MoO₄)₈:0.07 Pr^{3+} and Li₃Ba₂Gd₃ (MoO₄)₈: 0.08 Sm^{3+} are much comparable with that of conventional Y₂O₂S: Eu^{3+} red phosphor. This suggests that Li₃Ba₂Gd₃ (MoO₄)₈:0.07 Pr^{3+} and Li₃Ba₂Gd₃ (MoO₄)₈:0.08Sm^{3+} nanophosphors are suitable red-emitting phosphors in near UV / blue light GaN based White LEDs.

References

[1] S. T. Lai, S. Huang, W. M. Yen, Phys. Rev., B. 26, 2349 (1982).

[2] P. A. Rodnyi, P. Dorenbos, G.B. Stryganyuk, A. S. Voloshinovskii, A. S. Potapov, C.W.E. van Eijk, J. Phys. Condens. Matter.15, 719 (2003).

[3] R. F. Klevtsova, A. D. Vasil'ev, L. A. Glinskaya, A. D. Kruglik, N. M. Kozhevnikova, V.P. Korsun, Zh. Strukt. Khim. 33, 126 (1992).

[4] M. Rico, X. Han, C. Cascales, F. Esteban-Betegon and C. Zaldo, Opt. Express. 19, 7640 (2011) .

[5] M. Song, L. Zhang and G. Wang, J Alloys Compd, 480, 839 (2009).

[6] A. Garcia-Cortes and C. Cascales, Chem. Mater. 20, 3884 (2008).

[7] M. Song, L. Wang, N. Zhang, X. Tai and G. Wang, Materials.7, 496 (2014).

[8] Y. C. Chang, C. H. Liang, S. A. Yan and Y. S. Chang, J. Phys. Chem. C.114, 3645 (2010).

[9] Ye Jin, Jiahua Zhang, Shaozhe Lu, Haifeng Zhao, Xia Zhang and Xiaojun Wang, J. Phys. Chem. C. 112, 5860 (2008).

[10] H. Liang, Ye. Tao, Q. Su, S. Wang, J. Solid State Chem. 167, 435–440 (2002).

[11] Xuyong Yang, Jie Liu, Hong Yang, Xibin Yu, Yuzhu Guo,Yongqin Zhou and Jieyu Liu, J. Mater. Chem. 19, 3771 (2009).

[12] Enrico Cavalli, Fabio Angiuli, Philippe Boutinaud and Rachid Mahiou, J. Sol. Stat. Chem.185, 136-142 (2012).

[13] Xuyong Yang, Yongqin Zhou, Xibin Yu, Hilmi Volkan Demir and Xiao Wei Sun, J. Mater. Chem. 21, 9009 (2011).

[14] Xiang Lin, Xvsheng Qiao, Xianping Fan, Solid State Sciences. 13, 579 583 (2011)

[15] R.D. Shannon, Acta Crystallogr. 32, 751 (2012).

[16] R.P. Rao, J. Electrochem. Soc. 143, 189–197 (1996).

[17] A. Kato, S. Oishi, T. Shishido, M. Yamazaki, S. Lida, J. Phys. Chem. Solids, 66 2079 (2005).

[18] L. J. Burcham, I. E. Wachs, Spectrochim Acta A. 54 1355 (1998).

[19] M.H. Weng, R.Y. Yang, Y.M. Peng, J.L. Chen, Ceram. Int. 38, 1319 (2012).

[20] A. John Peter, I. B. Shameem Banu, Samuel Paul David, Thirumalai, J, J. Mater. Sci. Mater. Electron. 24, 4503–4509 (2013).

[21] Li-Ya Zhou, Jian-She Wei, Ling-Hong Yi, Fu-Zhong Gong, Jun-Li Huang and Wei Wang, Mater. Res. Bull. 44 1411 (2009).

[22] Z. H. Ju, R. P. Wei, J. X. Ma, C. R. Pang and W. S. Liu, J. Alloys Compd. 507, 133 (2010) .

[23] P.L. Li, Z.J. Wang, Z.P. Yang, Q.L. Guo, X. Li, Mater. Lett. 63, 751-753 (2009).

[24] QuhongZhang, JingWang, MeiZhang, WeijiaDing, QiangSu, J.Rare Earths, 24, 392 (2006).

[25] L. Y. Hu, H. W. Song, G. H. Pan, B.Yan, R. F.Qin,Q. L. Dai, L. B. Fan, S. W. Li, X. Bai, J. Lumin., 127, 371 (2007).12 Enrico Cavalli, Fabio Angiuli, Philippe Boutinaud and Rachid Mahiou, *J. Sol. Stat.Chem.*2012, **185,136**-142.

AUTHOR

YOUR KNOWLEDGE HAS VALUE

- We will publish your bachelor's and master's thesis, essays and papers

- Your own eBook and book - sold worldwide in all relevant shops

- Earn money with each sale

Upload your text at www.GRIN.com and publish for free